Addition & Subtr
Age 5-6

Alison Oliver

In a strange place, not too far from here, lives a scare of monsters.

A 'scare' is what some people call a group of monsters, but these monsters are really very friendly once you get to know them.

They are a curious bunch – they look very unusual, but they are quite like you and me, and they love learning new things and having fun.

In this book you will go on a learning journey with the monsters and you are sure to have lots of fun along the way.

Do not forget to visit our website to find out more about all the monsters and to send us photos of you in your monster mask or the monsters that you draw and make!

Contents

Sets of Mini-Monsters

Gran is a Monsterologist.
She has been to the wild wood to collect some mini-monsters.
Now she wants to add them together to see how many she has got in total.

Here are two groups of mini-monsters.
In maths, groups are called **sets**.

Things can be added together by counting all the objects in each set.

You just keep on counting and the last number is the **total**.
Total means **altogether**.

How many mini-monsters are there altogether?

Count   **1,2,3,4**   **5,6**   **Total is 6**.

1   Count the totals.
The last number in the count is the total.

a   b   c

6 moodles   7 squints   9 babbles

2 How many mini-monsters are there altogether?
= is used to mean **equals** or **is the same as**.

a 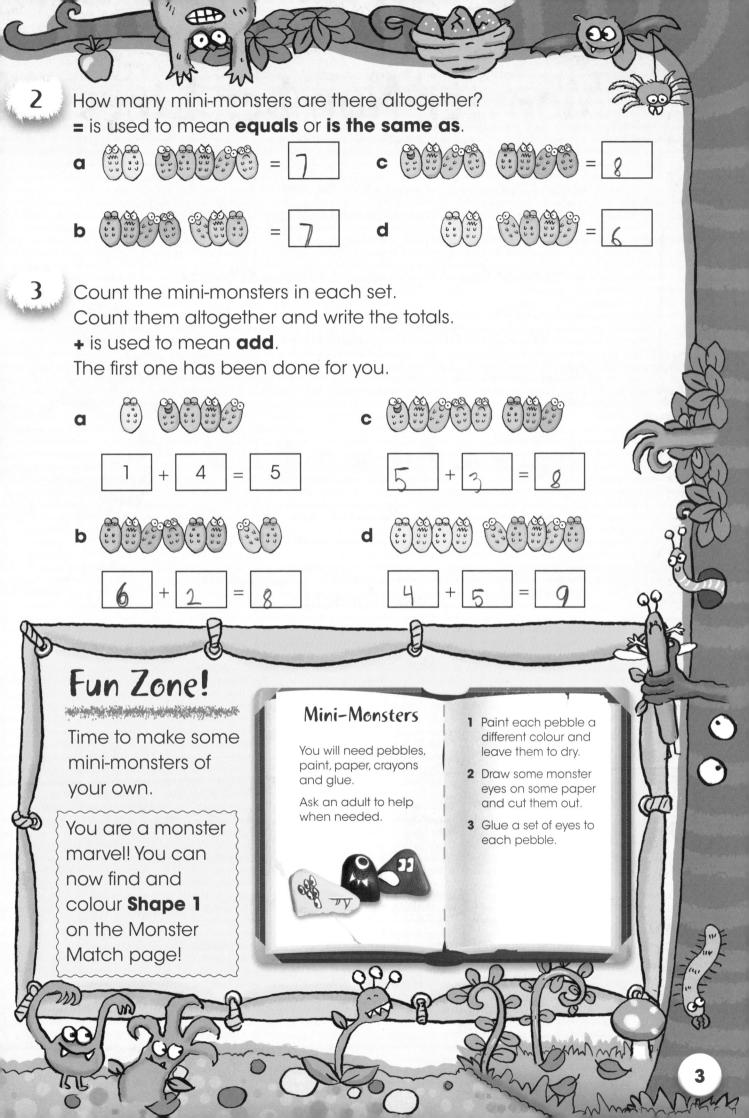 = 7 c = 8

b = 7 d = 6

3 Count the mini-monsters in each set.
Count them altogether and write the totals.
+ is used to mean **add**.
The first one has been done for you.

a

1 + 4 = 5

c

5 + 3 = 8

b

6 + 2 = 8

d

4 + 5 = 9

Fun Zone!

Time to make some mini-monsters of your own.

You are a monster marvel! You can now find and colour **Shape 1** on the Monster Match page!

Mini-Monsters

You will need pebbles, paint, paper, crayons and glue.

Ask an adult to help when needed.

1 Paint each pebble a different colour and leave them to dry.

2 Draw some monster eyes on some paper and cut them out.

3 Glue a set of eyes to each pebble.

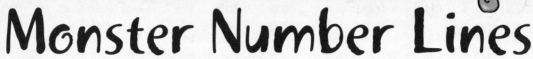

Monster Number Lines

Gran is teaching me to add like her.
She has made a number line out of monster footprints to help me.

A number line can be useful when **adding on**.

When adding sets of objects together we also use the words **plus** and **more than**.

What is 4 plus 2?
Work out 2 more than 4.

Count on the number line to check the answer
4 + 2 = 6 (this is called **a number sentence**).

Watch out for tricky sets with nothing in them.
Nothing is called **zero** or **nought**.
You stay still on the number line.

$$5 + \mathbf{0} = 5 \quad \text{and} \quad \mathbf{0} + 5 = 5$$

1 Find the totals for each of these.
Draw the jumps and write the answers.

a $3 + 5 = \boxed{8}$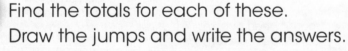

b $4 + 4 = \boxed{8}$

c $2 + 3 = \boxed{}$

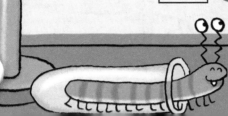

2 Now try these more difficult ones.

a $6 + 0 =$ ☐ 0 1 2 3 4 5 6 7 8 9 10

b $0 + 4 =$ ☐ 0 1 2 3 4 5 6 7 8 9 10

3 Use the number line to help answer these.

0 1 2 3 4 5 6 7 8 9 10

a $4 + 2 =$ ☐   **d** $6 + 2 =$ ☐   **g** 7 add 3 = ☐

b $4 + 4 =$ ☐   **e** $0 + 5 =$ ☐   **h** 3 plus 3 = ☐

c $2 + 3 =$ ☐   **f** $3 + 5 =$ ☐   **i** 5 more than 4 = ☐

Fun Zone!

Join the dots
to discover a
monster creature.

Congratulations!
You can now
find and colour
Shape 2 on the
Monster Match
page!

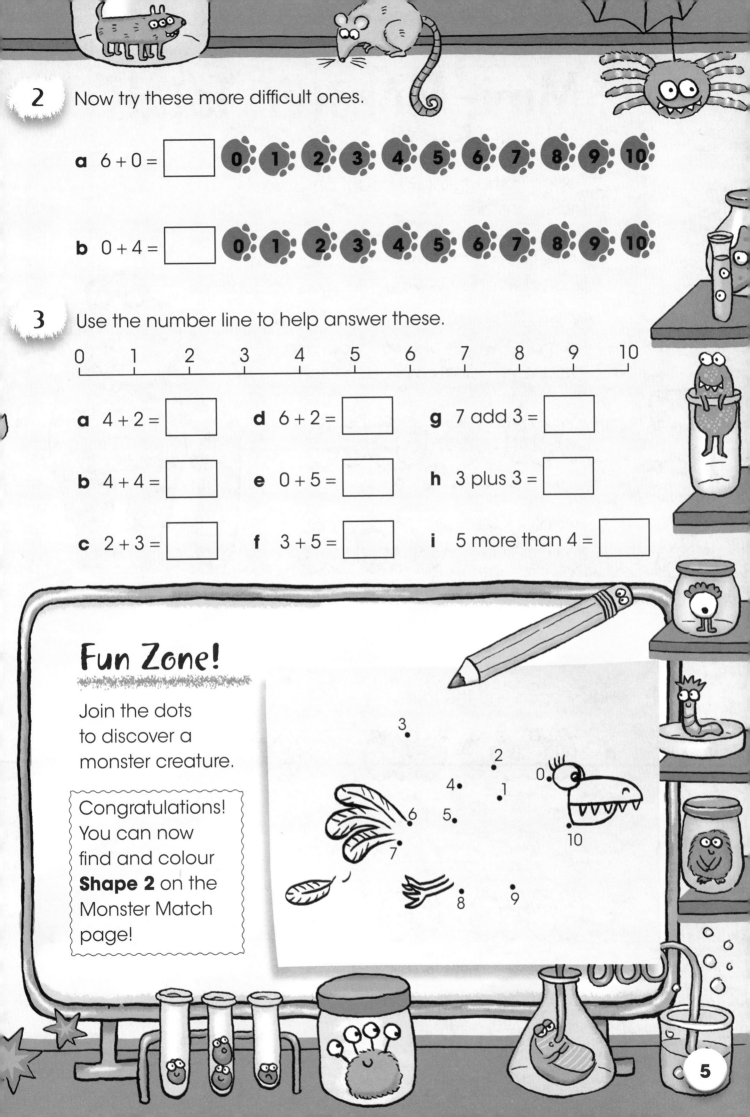

5

Mini-Monster Totals

Poggo is struggling to count all the mini-monsters.
They move around too much.
Sometimes a mini-monster moves from one set to another when Poggo's not looking!

When adding two numbers together, no matter which one you start with, the answer will always be the same.

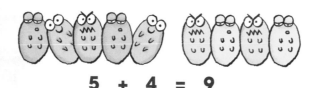

5 + 4 = 9

4 + 5 = 9

1 Help Poggo find the totals.

a

4 + 3 = ☐   3 + 4 = ☐

c

3 + 5 = ☐   5 + 3 = ☐

b

2 + 4 = ☐   4 + 2 = ☐

d

6 + 3 = ☐   3 + 6 = ☐

2 Count up these mini-monsters.
Write down your own number sentences and totals.
Remember to use **+** and **=**.

a

d

b

e

c

f

Fun Zone!

Make your own cress-head monster.

Watch your monster grow! You can now find and colour **Shape 3** on the Monster Match page!

Cress-Head Monster

You will need cress seeds, cotton wool, an old yoghurt pot, paint, paper, crayons and glue.

Ask an adult to help when needed.

1 Paint your old yoghurt pot and leave to dry.

2 Use crayons to draw some eyes on the paper. Cut them out and glue them onto the yoghurt pot.

3 Fill the yoghurt pot three quarters full of wet cotton wool.

4 Add the cress seeds on top of the cotton wool.

5 Keep the cotton wool wet and after a few days the cress will grow!

Monster Number Bonds

Gran now needs sets of mini-monsters that add up to 10 and 20.
Poggo will have to start counting.

$$6 + 4 = 10 \qquad 7 + 3 = 10 \qquad 10 + 0 = 10$$
$$16 + 4 = 20 \qquad 17 + 3 = 20 \qquad 20 + 0 = 20$$

Poggo can see a pattern!
A number with only one digit, such as 3 or 7, is called a **unit number**.
$7 + 3$ or $3 + 7$ always equal 10.
This is called a **number bond**.

So, $17 + 3 = 10 + 7 + 3$
$17 + 3 = 10 + 10$
$17 + 3 = 20$

1 Count the mini-monsters in each set and write the total.

a +  = $\Box$ + $\Box$ = $\Box$ moodles

 + 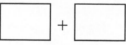 =  $\Box$ + $\Box$ = $\Box$ moodles

b + 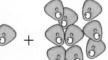 = $\Box$ + $\Box$ = $\Box$ squints

 + = $\Box$ + $\Box$ = $\Box$ squints

2 Poggo has found an animal while looking for mini-monsters.
Colour the shapes where the numbers add up to 20.

You can total numbers in any order.

0 1 2 3 4 5 6 7 8 9 10 11 12 13 14 15 16 17 18 19 20

Use the number line if you are not sure.
Try to start on the biggest number.

8 + 5	9 + 7
16 + 3	13 + 7
16 + 4	9 + 4
5 + 15	4 + 6
5 + 9	8 + 8
7 + 2	17 + 3
19 + 1	2 + 18
2 + 17	15 + 4
9 + 8	18 + 1
8 + 6	9 + 7

16 + 3 8 + 5 9 + 7 13 + 7 9 + 4 11 + 7
16 + 4 4 + 6
5 + 9 5 + 15 10 + 9 7 + 6
8 + 8 17 + 3 12 + 8
7 + 2 19 + 1 2 + 18 11 + 5 9 + 9
2 + 17 15 + 4 10 + 10 4 + 7
9 + 8 18 + 1 7 + 6 3 + 15
8 + 6 9 + 7 3 + 12 2 + 17 5 + 7

Which animal did Poggo find?

[   ]

Fun Zone!

Find each group of numbers shown below in the same order in the number search.

0	1	2	3
9	**10**	**11**	**12**
6	**7**	**8**	**9**
17	**18**	**19**	**20**

15	6	7	8	9
6	14	5	16	10
17	18	19	20	11
2	11	15	7	12
20	0	1	2	3

Well done! You can now find and colour **Shape 4** on the Monster Match page!

Monster Mechanics

Grandpa is a mechanic and needs to add up lots of objects to help him fix things. He often has to total three numbers to get the right number of bolts!

4 + 6 + 7

To make it easier, start with the largest number.

7 + 6 + 4 = 17

Look for pairs of numbers that are easy to total.

6 + 4 = 10

　　　　10 + 7 = 17

1 Write the number in each set.
Add these together to find the total.

a ☐ + ☐ = ☐

b ☐ + ☐ = ☐

c ☐ + ☐ = ☐

d ☐ + ☐ = ☐

2 Count the objects and add together three numbers.
Use your number bonds to help.

a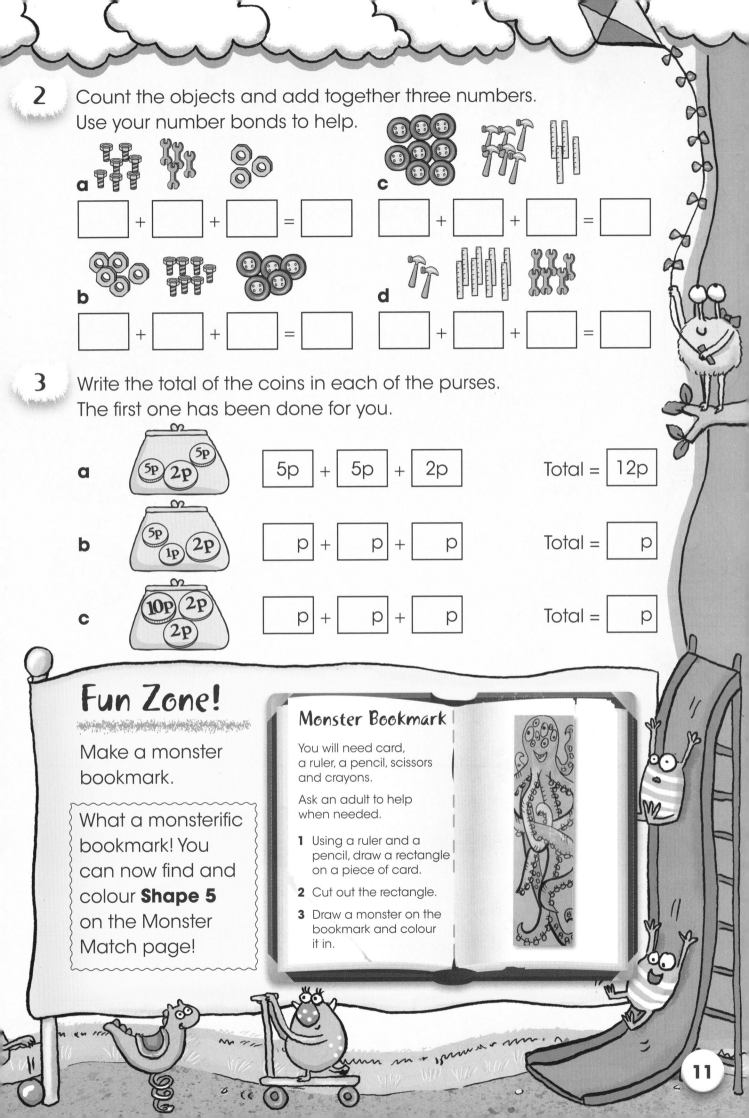

☐ + ☐ + ☐ = ☐

c

☐ + ☐ + ☐ = ☐

b

☐ + ☐ + ☐ = ☐

d

☐ + ☐ + ☐ = ☐

3 Write the total of the coins in each of the purses.
The first one has been done for you.

a | 5p | + | 5p | + | 2p | Total = | 12p |

b | p | + | p | + | p | Total = | p |

c | p | + | p | + | p | Total = | p |

Fun Zone!

Make a monster bookmark.

What a monsterific bookmark! You can now find and colour **Shape 5** on the Monster Match page!

Monster Bookmark

You will need card, a ruler, a pencil, scissors and crayons.

Ask an adult to help when needed.

1 Using a ruler and a pencil, draw a rectangle on a piece of card.

2 Cut out the rectangle.

3 Draw a monster on the bookmark and colour it in.

Missing Monster Maths

My name is Otto and I am helping Grandpa to fix Poggo's skateboard. Poggo uses it so much that it needs some new wheels. Grandpa is very forgetful and has lost the spare wheels! I'm helping him find them.

Remember your **number bonds**? These can help you find a number that is missing from a calculation.

? **+ 8 = 10**

What do you add to 8 to make 10? Start on 8 and count the jumps to 10.

```
0   1   2   3   4   5   6   7   8   9   10
```

2 + 8 = 10 (2 is the missing number)

1 Write the missing numbers on the wheels.

a 6 + ◯ = 10

d 1 + ◯ = 10

b ◯ + 5 = 10

e ◯ + 3 = 10

c 7 + ◯ = 10

f 10 + ◯ = 10

2 Draw more wheels to make each set total 20.
Write the missing numbers for each number sentence.
The first one has been done for you.

a 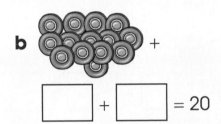 +

$$\boxed{13} + \boxed{7} = 20$$

c +

$$\boxed{} + \boxed{} = 20$$

b +

$$\boxed{} + \boxed{} = 20$$

d +

$$\boxed{} + \boxed{} = 20$$

3 Draw a line to join each number sentence to the correct missing number.

$$\boxed{} + 11 = 20 \qquad \boxed{} + 7 = 10 \qquad \boxed{} + 14 = 20$$

5   3   9   6   7   8

$$5 + \boxed{} = 10 \qquad 5 + \boxed{} = 13 \qquad \boxed{} + 8 = 15$$

Fun Zone!

Here is a monster picture. Colour each number a different colour using the colour code.

Well done! You can now find and colour **Shape 6** on the Monster Match page!

Code: 1 = blue   4 = red
2 = green   5 = purple
3 = yellow   6 = orange

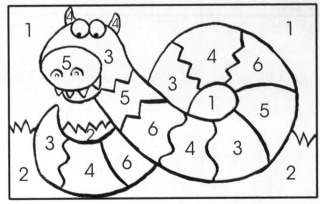

Monster Challenge 1

1 Count each set and write the totals.

a 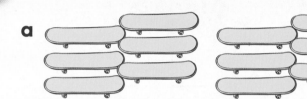  [ ] + [ ] = [ ]

b 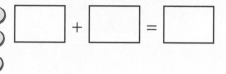 [ ] + [ ] = [ ]

2 Write the totals for each of these.

a $6 + 8 =$ [ ]   **h** $9 + 3 =$ [ ]

b $4 + 9 =$ [ ]   **i** $7 + 7 =$ [ ]

c $5 + 8 =$ [ ]   **j** $6 + 9 =$ [ ]

d $6 + 7 =$ [ ]   **k** $5 + 6 =$ [ ]

e $11 + 4 =$ [ ]   **l** $0 + 7 =$ [ ]

f $13 + 2 =$ [ ]   **m** $3 + 5 =$ [ ]

g $14 + 0 =$ [ ]   **n** $2 + 15 =$ [ ]

3 Match the hats with the same totals by colouring them the same colour.

13 + 7   12 + 4   11 + 9

9 + 7   14 + 6   14 + 2

4 Complete these addition squares.
For each square, add the top number to the far left number.
An example has been done for you (7 + 3 = 10).

a

+	8	7	6
4			
3		10	
7			

b

+	4	9	8
7			
6			
5	9		

5 Poggo has been jumping in the mud.
Write a number sentence to match Poggo's jumps on each number line.

a   12 13 14 15 16 17 18 19 20

12 + ☐ + ☐ = ☐

b   11 12 13 14 15 16 17 18 19 20

11 + ☐ + ☐ = ☐

Taking Away

When some items in a set are **taken away**, the total is less than the number at the start. Try this by crossing out the number of items to take away.

Poggo is hungry so he is having a break from counting.
He is sharing some cakes with the other monsters.

How many cakes are there? **6**
We take away two **– 2**

Cross out the cakes to take them away.

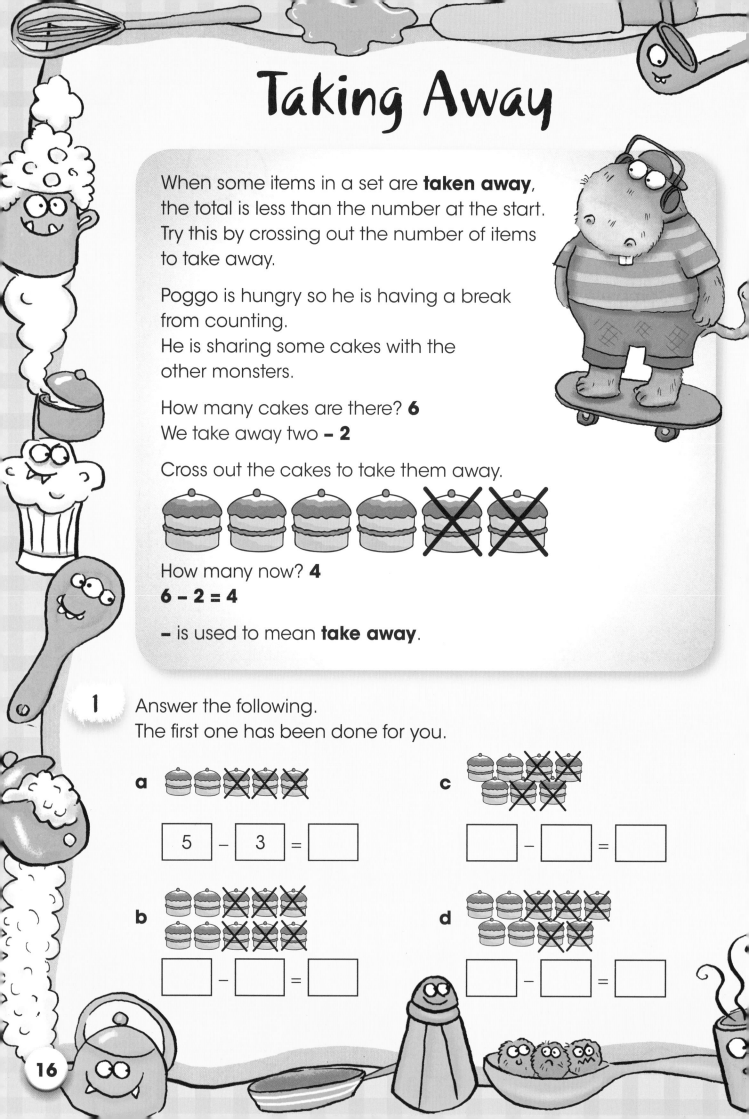

How many now? **4**
6 – 2 = 4

– is used to mean **take away**.

1 Answer the following.
The first one has been done for you.

a

| 5 | – | 3 | = | |

b

| | – | | = | |

c

| | – | | = | |

d

| | – | | = | |

2 Another word for take away is **subtract**.
Complete these.

a 10 take away 7 is ☐

10 – 7 = ☐

b 8 subtract 4 is ☐

8 – 4 = ☐

c 9 subtract 4 is ☐

9 – 4 = ☐

d 5 take away 2 is ☐

5 – 2 = ☐

3 Draw a line to join each plate to the cake
with the correct answer.

10 – 9   9 – 4   8 – 1   7 – 5

7   2   1   5   3   4   0   1

6 – 3   4 – 0   3 – 3   5 – 4

Fun Zone!

Time to make
some little
monsters.

Terrific monsters!
You can now
find and colour
Shape 7 on
the Monster
Match page!

Little Monsters

You will need paper,
crayons, white paint and
a black pencil.

Ask an adult to help
when needed.

1 Use crayons to draw
colourful circles on
some paper. Leave
space between
the circles.

2 Paint white spots for
the monsters' eyes.
Leave to dry.

3 Use a pencil to draw
arms, legs, teeth and
any other features
you would like your
monsters to have.

Subtraction Facts to 10

Poggo loves playing frisbee with his pet Zak.
Zak is very good at frisbee.
He can even throw it back for Poggo to catch!

Number lines can help you count back when you are taking away.

Count the jumps.
Start at 7 and count back 4.

0 1 2 3 4 5 6 7 8 9 10

7 – 4 = 3

1 Use these number lines to count back and write the answers.
The first one has been done for you.

a 8 – 5 = [3]   **0 1 2 3 4 5 6 7 8 9 10**

b 6 – 3 = [ ]   **0 1 2 3 4 5 6 7 8 9 10**

c 7 – 5 = [ ]   **0 1 2 3 4 5 6 7 8 9 10**

d 9 – 9 = [ ]   **0 1 2 3 4 5 6 7 8 9 10**

2 Use the number line to help you answer these.

0 1 2 3 4 5 6 7 8 9 10

a 5 − 3 = ☐

c 7 − 7 = ☐

e 10 − 10 = ☐

b 7 − 4 = ☐

d 9 − 4 = ☐

f 8 − 0 = ☐

3 Draw a line to join each question to the correct answer.

5 − 5   8 − 4   9 − 2   10 − 5

0 1 2 3 4 5 6 7 8 9 10

7 − 4   9 − 3   6 − 5   10 − 0

Fun Zone!

Join the dots to discover a monster pet.

Congratulations! You can now find and colour **Shape 8** on the Monster Match page!

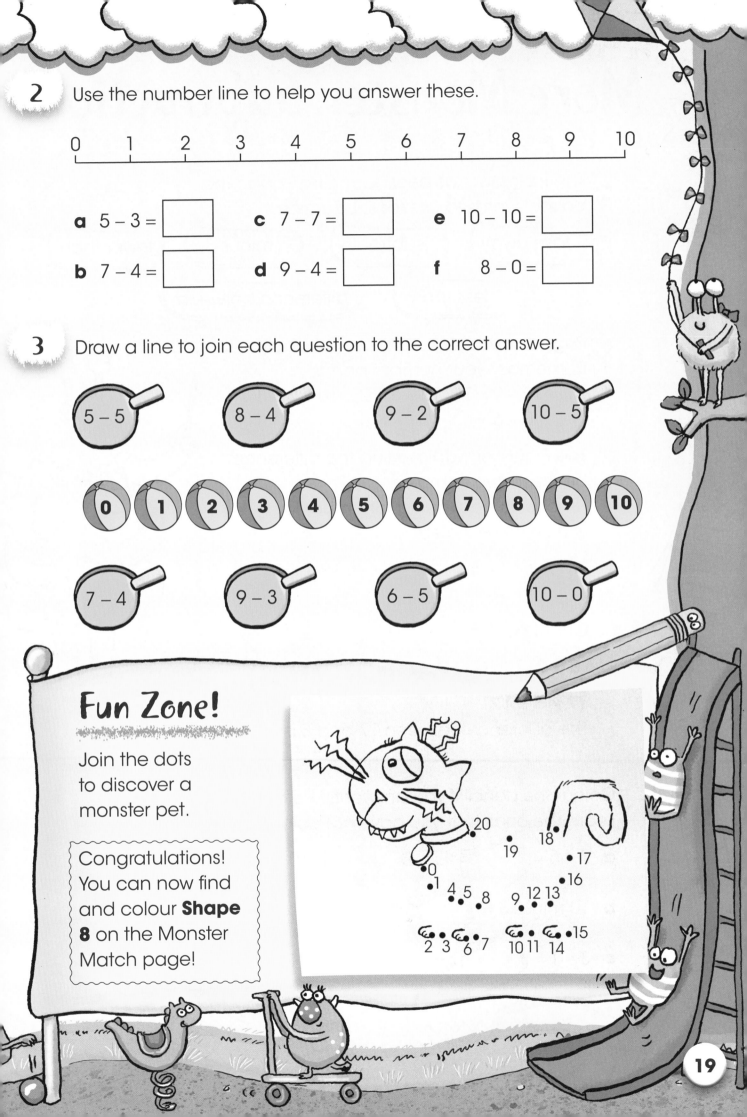

More Monster Subtraction

The Professor has been teaching Poggo the correct words to use for subtraction.

> take away   subtract   minus   fewer than
>
> less than   difference between

Poggo needs to learn these subtraction facts too.
Remember your number bonds!

$$10 - 9 = 1 \quad 10 - 8 = 2 \quad 10 - 7 = 3$$
$$10 - 6 = 4 \quad 10 - 5 = 5$$

The Professor wants to find the difference between two numbers.
He counts back using a number line written on his notepads.

What is $17 - 5$?

Start at 17.
Count back 5.

8   9   10   11   12   13   14   15   16   17   18   19   20

$17 - 5 = 12$

The difference between 17 and 5 is 12.

1   Colour the pencil if you answer these quickly in your head.
Learn the ones that you have not coloured.

a $7 - 5 =$

b   10 minus 3 =

c $3 - 1 =$

d $10 - 5 =$

e   4 fewer than 9 =

f   5 less than 6 =

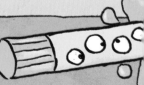

2 Draw the jumps on the number lines.
Write the answer for each of these.

a 14 – 10 = ☐   3 4 5 6 7 8 9 10 11 12 13 14 15 16 17

b 18 – 5 = ☐   9 10 11 12 13 14 15 16 17 18 19 20

c 11 – 11 = ☐   0 1 2 3 4 5 6 7 8 9 10 11 12 13

d 7 – 0 = ☐   3 4 5 6 7 8 9 10 11 12 13 14 15 16

3 Use the number line to help work out each of these.

0 1 2 3 4 5 6 7 8 9 10 11 12 13 14 15 16 17 18 19 20

a 9 – 5 = ☐   **c** 12 – 12 = ☐   **e** 14 – 5 = ☐

b 12 – 4 = ☐   **d** 13 – 8 = ☐   **f** 17 – 0 = ☐

Fun Zone!

Make some messy monster handprints!

Great monster handprints! You can now find and colour **Shape 9** on the Monster Match page!

Handprint Monsters

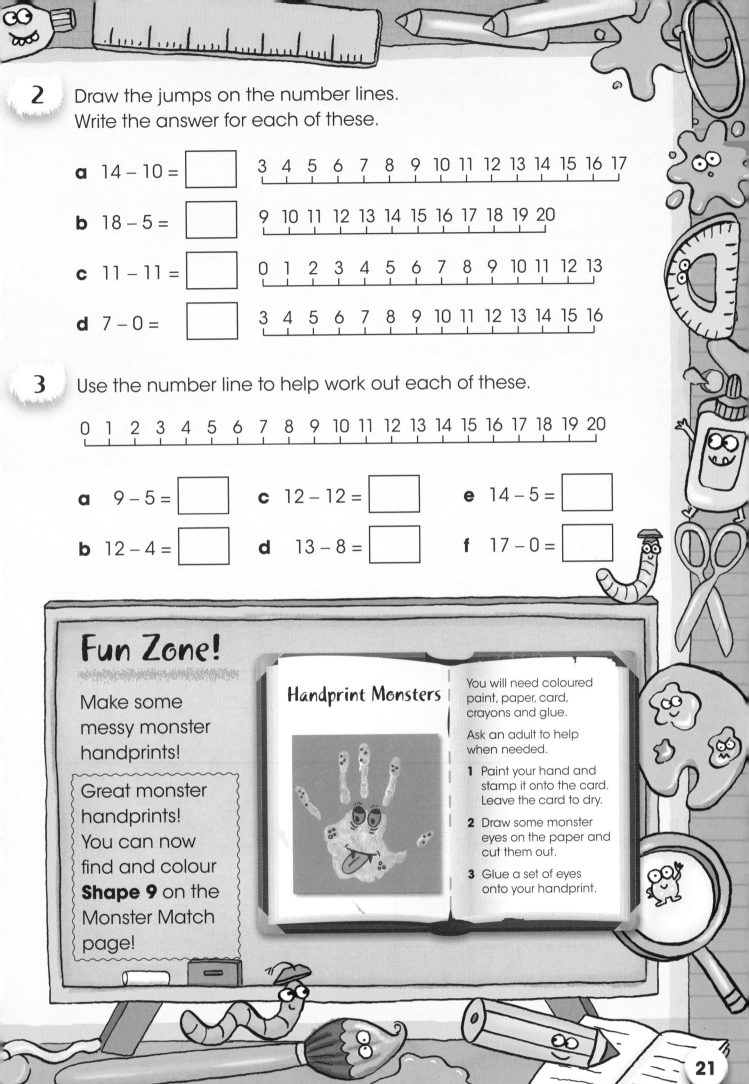

You will need coloured paint, paper, card, crayons and glue.

Ask an adult to help when needed.

1 Paint your hand and stamp it onto the card. Leave the card to dry.

2 Draw some monster eyes on the paper and cut them out.

3 Glue a set of eyes onto your handprint.

Tricky Trios

I am very good at counting now!
Dad has found sets of crystals in the
monster mines.
He has asked me to count these sets of crystals.

A trio is a set of **three numbers** that make
addition and **subtraction families**.
These are useful for learning number facts.

 4   2   6

$$4 + 2 = 6 \qquad\qquad 2 = 6 - 4$$
$$2 + 4 = 6 \qquad\qquad 4 = 6 - 2$$

1   Help Poggo write the addition and subtraction facts from each
set of crystals.

a   9   7   2

| 7 | + | ☐ | = | ☐ | | ☐ | = | ☐ | − | 7 |

| 2 | + | ☐ | = | ☐ | | ☐ | = | ☐ | − | 2 |

b   5   8   3

| ☐ | + | ☐ | = | 8 | | 8 | − | ☐ | = | ☐ |

| ☐ | + | ☐ | = | 8 | | 8 | − | ☐ | = | ☐ |

c   4   3    7

| ☐ | + | ☐ | = | ☐ | | ☐ | − | ☐ | = | ☐ |

| ☐ | + | ☐ | = | ☐ | | ☐ | − | ☐ | = | ☐ |

2 Write the missing addition and subtraction facts for each set of crystals.

a 9 15 6

[ ] + [ ] = 15   15 – [ ] = [ ]

[ ] + [ ] = 15   15 – [ ] = [ ]

b 12 5 7

5 + [ ] = [ ]   [ ] = 12 – 7

[ ] + 5 = [ ]   7 = [ ] – [ ]

3 Write two number sentences for each picture.

a

7 + [ ] = [ ]

13 – [ ] = [ ]

b

8 + [ ] = [ ]

15 – [ ] = [ ]

Fun Zone!

Find five differences between these two pictures of Poggo.

Well done! You can now find and colour **Shape 10** on the Monster Match page!

Missing Monster Numbers

Dad has lost some of his crystals!
He is using the trio families to work out how many are missing.

? take away 5 = 8.

You know 8 add 5 = 13, so ? must be 13.

You can use a number line to check your answer.

5 6 7 8 9 10 11 12 13 14 15

---

1 Use the number lines to find the missing numbers.
The first one has been done for you.

a ? – 5 = 6   0 1 2 3 4 5 6 7 8 9 10 11 12 13 14 15 16 17 18 19 20

$\boxed{6}$ + $\boxed{5}$ = $\boxed{11}$? = $\boxed{11}$

b ? – 3 = 12   0 1 2 3 4 5 6 7 8 9 10 11 12 13 14 15 16 17 18 19 20

$\boxed{12}$ + $\boxed{3}$ = $\boxed{}$? = $\boxed{}$

c ? – 8 = 7   0 1 2 3 4 5 6 7 8 9 10 11 12 13 14 15 16 17 18 19 20

$\boxed{}$ + $\boxed{7}$ = $\boxed{}$? = $\boxed{}$

d ? – 5 = 4   0 1 2 3 4 5 6 7 8 9 10 11 12 13 14 15 16 17 18 19 20

$\boxed{}$ + $\boxed{}$ = $\boxed{}$? = $\boxed{}$

2 This monsterbot takes away 5 from all numbers.
The numbers going in have been lost.
Work out what they should be.

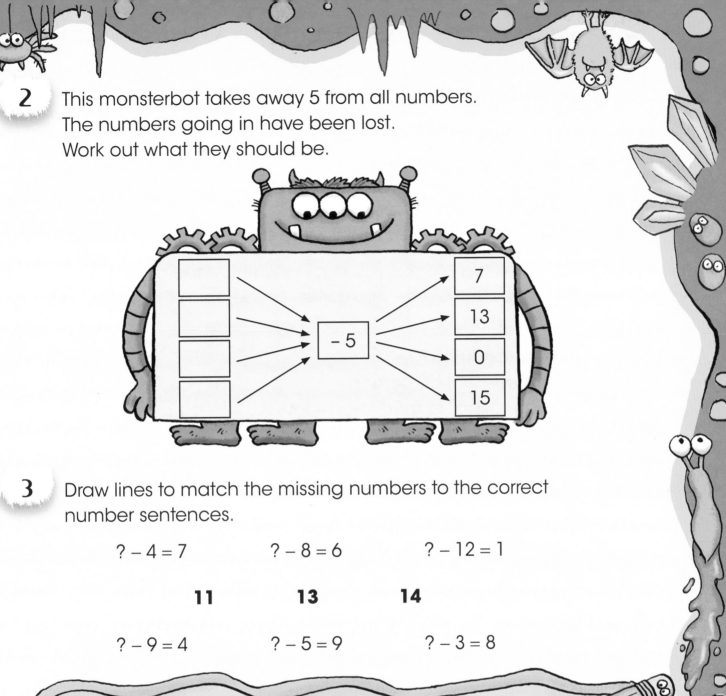

3 Draw lines to match the missing numbers to the correct number sentences.

? – 4 = 7   ? – 8 = 6   ? – 12 = 1

11   **13**   **14**

? – 9 = 4   ? – 5 = 9   ? – 3 = 8

Fun Zone!

Use crayons to colour in Zak.

A marvellous monster picture! You can now find and colour **Shape 11** on the Monster Match page!

Monster Problem Solving

Poggo is collecting Nano's toys from two rooms in their home.
He puts them in two piles.
Now he wonders how many there are altogether.

Problems are easier if you decide whether the answer is going to be a bigger or smaller number than the one you started with.

Is it an addition?
The answer will be bigger.
In the living room there are 14 toys and in the kitchen there are 8 toys.
Altogether there are $14 + 8 = 22$ toys.

Is it a take away?
The answer will be smaller.
How many more toys are there in the living room than in the kitchen?
The difference is $14 - 8 = 6$ toys.

1 Mum has 20p to spend at the shop on a new rattle for Nano.
The shopkeeper works out the difference
and gives her the change.
Write the change from 20p for each of these rattles.
The first one has been done for you.

a (14p)

20p – $\boxed{14p}$ = $\boxed{6p}$ change

c (11p)

20p – $\boxed{p}$ = $\boxed{p}$ change

b (15p)

20p – $\boxed{p}$ = $\boxed{p}$ change

d (18p)

20p – $\boxed{p}$ = $\boxed{p}$ change

2 Nano has hidden some of his toys.
Can you work out how many he has hidden?
Write a number sentence for each.

Nano started with 18 toys…

a Now he has 13 | 18 | − | 13 | = | |

b Now he has 9 | | − | | = | |

c Now he has 0 | | − | | = | |

d Now he has 3 | | − | | = | |

3 Poggo is in the toy shop.
Help him work out how many toys are on each shelf.

a 2 rattles + 15 teddy bears = [ ] toys

b 8 dolls + 9 yo-yos = [ ] toys

c 6 robots + 12 trains = [ ] toys

d 7 yo-yos + 11 trains = [ ] toys

Fun Zone!

Follow the squares
and fill in the missing
numbers to help
Nano reach his rattle.

Well done! You
can now find and
colour **Shape 12**
on the Monster
Match page!

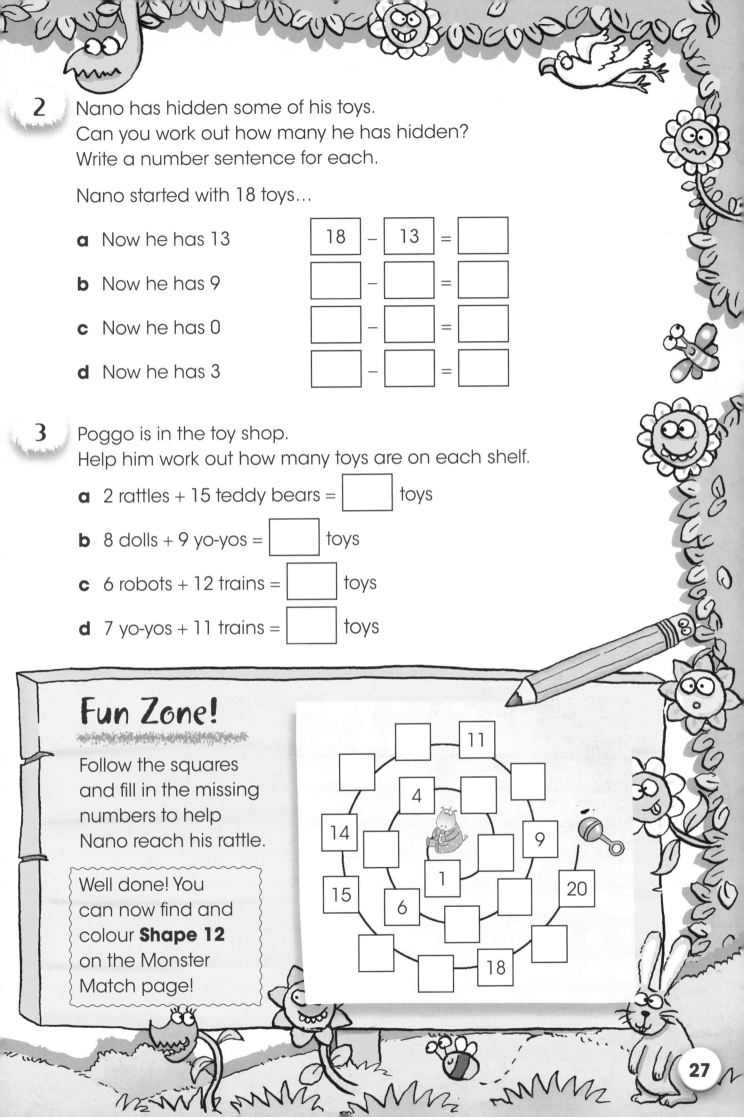

Monster Challenge 2

1 Look at each number line and write the missing numbers in each subtraction.

a 8 9 10 11 12 13 14 15 16 17 18 19 20

[17] – [　　] = [　　]

b 5 6 7 8 9 10 11 12 13 14 15

[14] – [　　] = [　　]

2 Write the numbers coming out of the monsterbot.

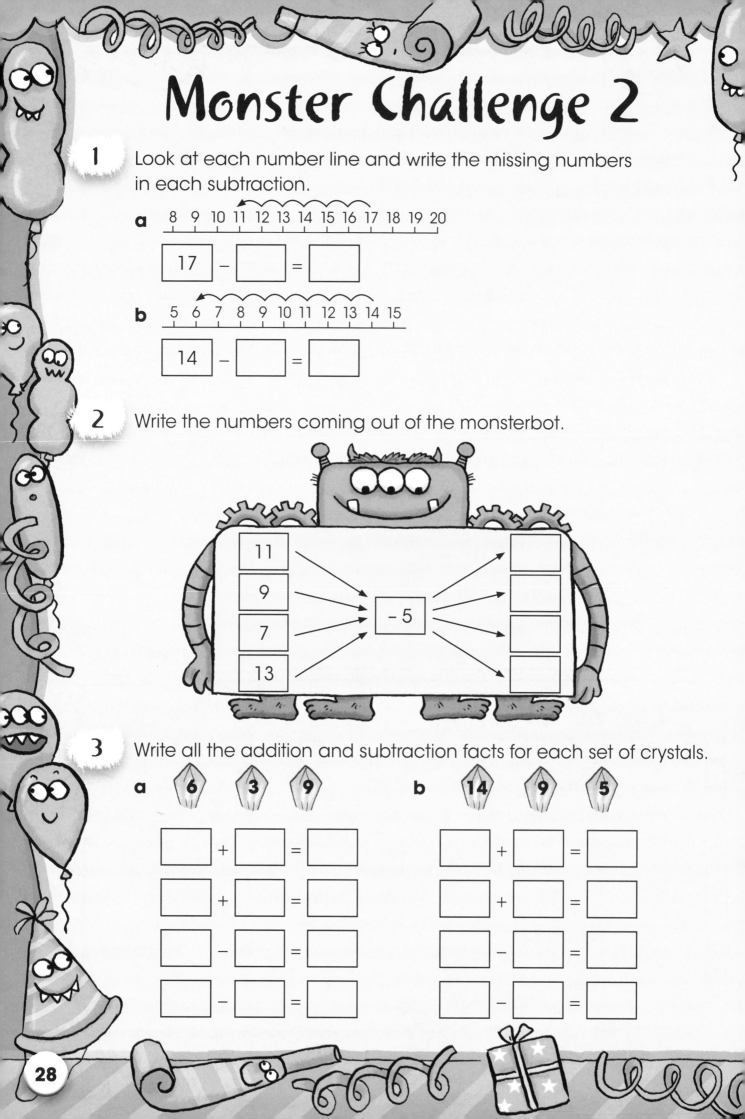

[11]
[9]
[7]
[13]

– 5

3 Write all the addition and subtraction facts for each set of crystals.

a 6   3   9

[　　] + [　　] = [　　]

[　　] + [　　] = [　　]

[　　] – [　　] = [　　]

[　　] – [　　] = [　　]

b 14   9   5

[　　] + [　　] = [　　]

[　　] + [　　] = [　　]

[　　] – [　　] = [　　]

[　　] – [　　] = [　　]

4 Write the difference between the numbers on each pair of skateboards.

a 12   8

☐ – ☐ = ☐

c 16   7

☐ – ☐ = ☐

b 13   6

☐ – ☐ = ☐

d 18   5

☐ – ☐ = ☐

5 Write the missing numbers for each of these.

a 17 – 3 = ☐

c 11 – 4 = ☐

b ☐ – 2 = 4

d 13 – ☐ = 8

6 Answer the following questions.

a I am thinking of a number.
When I subtract 3 from this number, I am left with 7.

Which number am I thinking of? ☐

b I am thinking of a number.
It is 5 less than 12.

Which number am I thinking of? ☐

I knew you could do it!
You have made it to the end of the book.
You are a magnificent monster!

Answers

Page 2
1 a 6   **b** 7   **c** 9

Page 3
2 a 7   **b** 7   **c** 8   **d** 6
3 b $6 + 2 = 8$ **c** $5 + 3 = 8$ **d** $4 + 5 = 9$

Page 4
1 a 8
   b 8   0 1 2 3 4 5 6 7 8 9 10
   c 5   0 1 2 3 4 5 6 7 8 9 10

Page 5
2 a 6
   b 4   0 1 2 3 4 5 6 7 8 9 10
3 a 6   **c** 5   **e** 5   **g** 10   **i** 9
   b 8   **d** 8   **f** 8   **h** 6

Fun Zone

Page 6
1 a $4 + 3 = 7; 3 + 4 = 7$ **c** $3 + 5 = 8; 5 + 3 = 8$
   b $2 + 4 = 6; 4 + 2 = 6$ **d** $6 + 3 = 9; 3 + 6 = 9$

Page 7
2 a $2 + 3 = 5$ **c** $3 + 1 = 4$ **e** $2 + 6 = 8$
   b $4 + 2 = 6$ **d** $3 + 5 = 8$ **f** $5 + 2 = 7$

Page 8
1 a $2 + 8 = 10$ **b** $1 + 9 = 10$
    $12 + 8 = 20$ $11 + 9 = 20$

Page 9
2

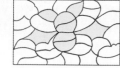

Animal: Bat

Fun Zone

15	6	7	8	9
6	14	5	16	10
17	18	19	20	11
2	11	15	7	12
20	0	1	2	3

Page 10
1 a $4 + 9 = 13$ **c** $3 + 8 = 11$
   b $5 + 7 = 12$ **d** $10 + 6 = 16$

Page 11
2 a $7 + 4 + 3 = 14$ **c** $8 + 6 + 4 = 18$
   b $5 + 7 + 5 = 17$ **d** $2 + 8 + 6 = 16$
3 b $5p + 2p + 1p = 8p$ **c** $10p + 2p + 2p = 14p$

Page 12
1 a 4   **b** 5   **c** 3   **d** 9   **e** 7   **f** 0

Page 13
2 b $14 + 6 = 20$
   c
    $10 + 10 = 20$
   d $16 + 4 = 20$
3 $9 + 11 = 20; 3 + 7 = 10; 6 + 14 = 20;$
   $5 + 5 = 10; 5 + 8 = 13; 7 + 8 = 15$

Fun Zone

Page 14
1 a $6 + 6 = 12$ **b** $5 + 8 = 13$
2 a 14   **d** 13   **g** 14   **j** 15   **m** 8
   b 13   **e** 15   **h** 12   **k** 11   **n** 17
   c 13   **f** 15   **i** 14   **l** 7

Page 15
3 $13 + 7$ matches $14 + 6$ and $11 + 9$, answer $= 20$
   $14 + 2$ matches $9 + 7$ and $12 + 4$, answer $= 16$

4

+	8	7	6
4	**12**	**11**	**10**
3	**11**	10	**9**
7	**15**	**14**	**13**

+	4	9	8
7	**11**	**16**	**15**
6	**10**	**15**	**14**
5	9	**14**	**13**

5 a $12 + 3 + 2 = 17$ **b** $11 + 5 + 4 = 20$

Page 16
1 b $10 - 6 = 4$ **c** $7 - 4 = 3$ **d** $9 - 5 = 4$

Page 17
2 a 7; 7   **b** 4; 4   **c** 5; 5   **d** 3; 3
3

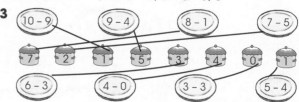

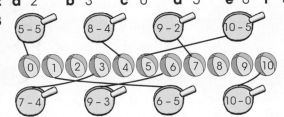

Page 18

1 b 3   **c** 2   **d** 0

Page 19

2 a 2   **b** 3   **c** 0   **d** 5   **e** 0   **f** 8

3

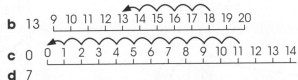

Fun Zone

Page 20

1 a 2   **b** 7   **c** 2   **d** 5   **e** 5   **f** 1

Page 21

2 a 4   3 4 5 6 7 8 9 10 11 12 13 14 15 16 17

b 13   9 10 11 12 13 14 15 16 17 18 19 20

c 0   0 1 2 3 4 5 6 7 8 9 10 11 12 13 14

d 7

3 a 4   **b** 8   **c** 0   **d** 5   **e** 9   **f** 17

Page 22

1 The answers can be shown in any order.

   a $7 + 2 = 9$ $2 = 9 - 7$ $2 + 7 = 9$ $7 = 9 - 2$

   b $5 + 3 = 8$ $8 - 5 = 3$ $3 + 5 = 8$ $8 - 3 = 5$

   c $4 + 3 = 7$ $7 - 3 = 4$ $3 + 4 = 7$ $7 - 4 = 3$

Page 23

2 a $9 + 6 = 15$ $15 - 9 = 6$

     $6 + 9 = 15$ $15 - 6 = 9$

   b $5 + 7 = 12$ $5 = 12 - 7$

     $7 + 5 = 12$ $7 = 12 - 5$

3 a $7 + 6 = 13$

     $13 - 7 = 6$ or $13 - 6 = 7$

   b $8 + 7 = 15$

     $15 - 7 = 8$ or $15 - 8 = 7$

Fun Zone

Page 24

1 b 0 1 2 3 4 5 6 7 8 9 10 11 12 13 14 15 16 17 18 19 20

     $12 + 3 = 15$ $? = 15$

   c 0 1 2 3 4 5 6 7 8 9 10 11 12 13 14 15 16 17 18 19 20

     $8 + 7 = 15$ $? = 15$

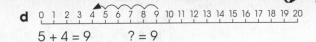

d 0 1 2 3 4 5 6 7 8 9 10 11 12 13 14 15 16 17 18 19 20

     $5 + 4 = 9$ $? = 9$

Page 25

2

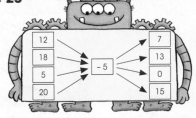

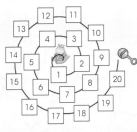

3 $? - 4 = 7$ $? - 8 = 6$ $? - 12 = 1$

     11   **13**   **14**

   $? - 9 = 4$ $? - 5 = 9$ $? - 3 = 8$

Page 26

1 b $20p - 15p = 5p$ change

   c $20p - 11p = 9p$ change

   d $20p - 18p = 2p$ change

Page 27

2 a $18 - 13 = 5$ **c** $18 - 0 = 18$

   b $18 - 9 = 9$ **d** $18 - 3 = 15$

3 a $2 + 15 = 17$ **c** $6 + 12 = 18$

   b $8 + 9 = 17$ **d** $7 + 11 = 18$

Fun Zone

Page 28

1 a $17 - 6 = 11$ **b** $14 - 8 = 6$

2

3 The answers can be shown in any order.

   a $6 + 3 = 9$ $3 + 6 = 9$ $9 - 6 = 3$ $9 - 3 = 6$

   b $9 + 5 = 14$ $5 + 9 = 14$ $14 - 9 = 5$ $14 - 5 = 9$

Page 29

4 a $12 - 8 = 4$ **c** $16 - 7 = 9$

   b $13 - 6 = 7$ **d** $18 - 5 = 13$

5 a 4   **b** 6   **c** 7   **d** 5

6 a 10 $(10 - 3 = 7)$ **b** 7 $(12 - 5 = 7)$

Monster Match

Each time you complete a topic in this book, you will be awarded a shape number.

Find and colour the shapes in the picture of Poggo that match the numbers you have been given.

As you work through the book you will gradually see Poggo come to life!